ESL For Farm Safety

AFOP's Working with English Series

Student Workbook

Acknowledgements

Special thanks to:

Cathy Kronopolus
Chief, Certification and Occupational Safety Branch
US Environmental Protection Agency, Office of Pesticide Programs

Carol Parker
Project Officer
EPA/OPP

Kevin Keany
Deputy Chief, Certification and Occupational Safety Branch
EPA/OPP

John Leahy
Environmental Protection Specialist
Health Effects Division
EPA/OPP

ISBN 1-886567-06-9

Table of Contents

1 Lesson One

Introductions

What's your name? Where are you from?

My name is Enrique. I'm from Mexico.

What's your name?

Name ____

First Last

Where are you from?

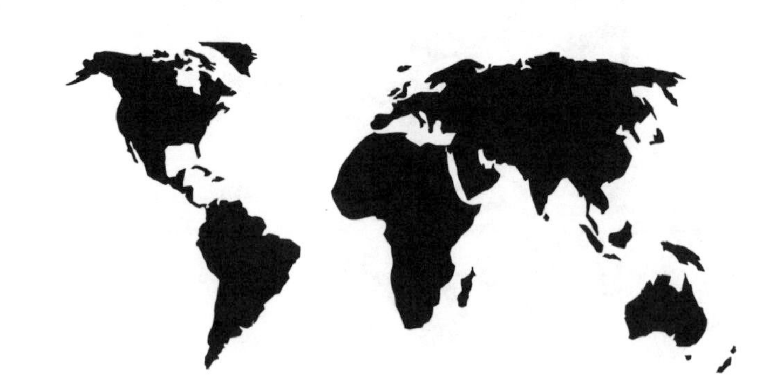

I'm from ____________________.

Where do you live?

Name	______________		
	First		Last
Address	______________		
	Street		
	______________		
	City	State	Zip Code

☎ My telephone number is ______________ .

Are you married or single?

Enrique and Blanca

married

Alma

single

Ask your classmates.

Name	Married	Single
Enrique	X	
Blanca	X	
Alma		X

Do you have any children?

Yes, I do. I have one son and one daughter.

Ask your classmates.

How many?

Name	Children	Sons	Daughters

Do you have children?

No, I don't.

Where do you work?

I work on the farm.

on the farm

in the fields

in the greenhouse

in the orchards

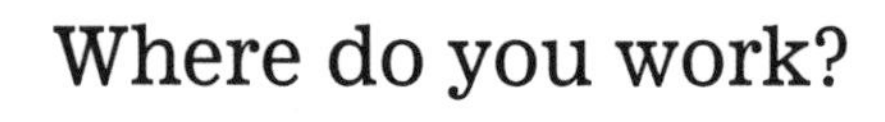

Where do you work?

I work ______________________.

What do you do?

I pick apples.

What do you do? I ____________________.

Which vegetables or fruits do you work with?

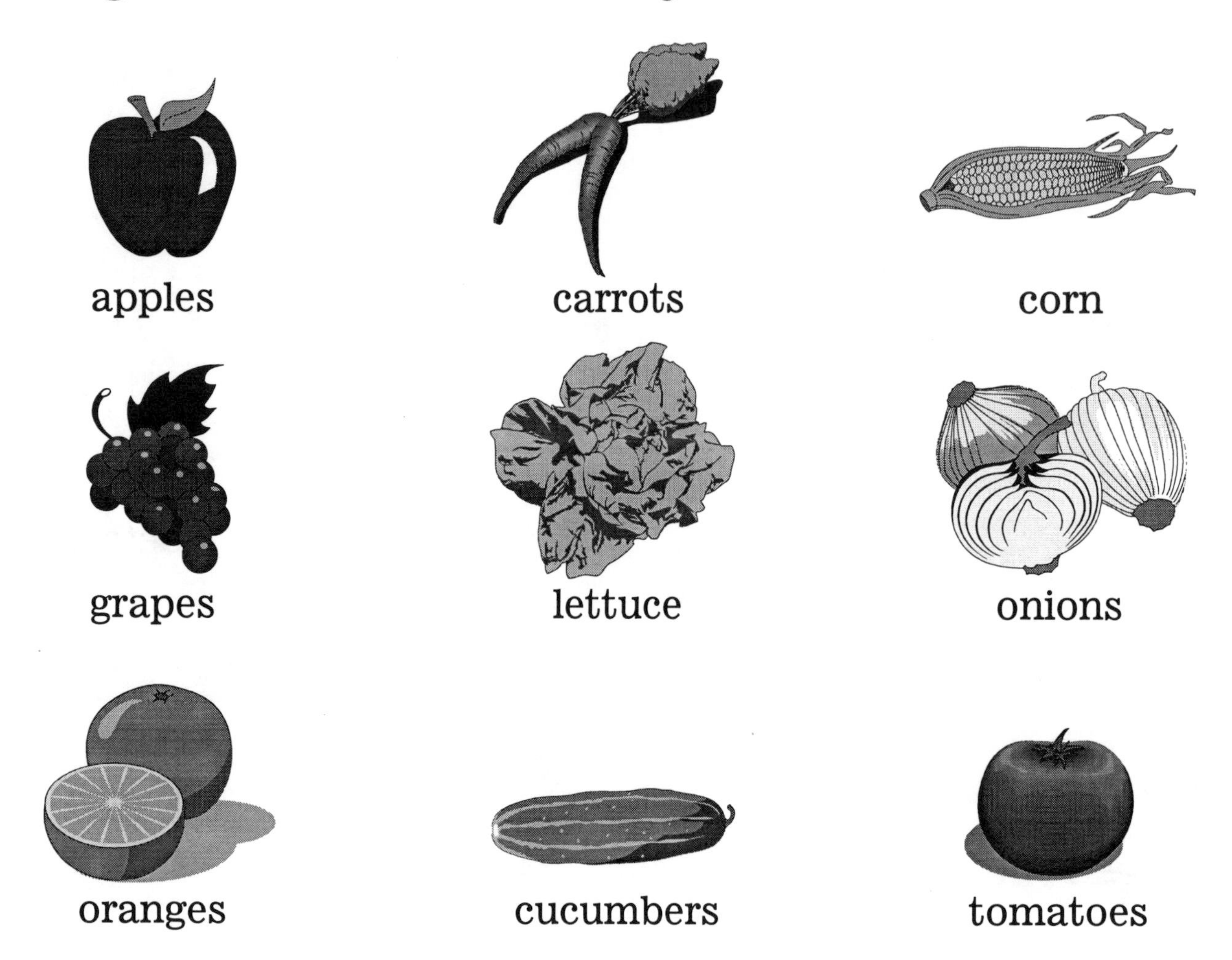

I work with ______________________.

Enrique's Story

Enrique Torres was born in Mexico. Now he lives at 44 Allen Lane in Belle Glade, Florida. He is married to Blanca. They have 3 sons and 2 daughters. Enrique works in the fields. He picks tomatoes.

1. What is Enrique's last name?
2. Where is he from?
3. Where does Enrique live?
4. Is he married or single?
5. Does Enrique have any children?
6. Where does Enrique work?
7. What does he do?

Lesson 1
Review

Match the words with the pictures.

____

____

____

____

____

____

prune orchards fields

pick sort greenhouse

Fill in the blanks.

1. My first name is________________________________.
2. My last name is________________________________.
3. My address is________________________________

   ____.
4. My telephone number is__________________________.
5. My zip code is________________________________.
6. My city is___________________________________.
7. My state is__________________________________.
8. My country is________________________________.

2 Lesson Two

What are pesticides?

Pesticides can control pests and diseases.

Insects

Weeds

Mold (Fungi)

Rodents

There are other types of pests too. What pests do you see at work?

1. ____________ 2. ____________ 3. ____________

Pesticides are poison. "-cide" means kill.

Pesticides can kill insects, rodents, weeds, mold(fungi), and other diseases that kill plants.

Pesticides kill pests.

Insecticides kill _______________.

Herbicides kill _______________.

Fungicides kill _______________.

Rodenticides kill _______________.

Pesticides are dangerous.
Pesticides can hurt you.
Sometimes pesticides can kill people.

Look at the warning on the label.

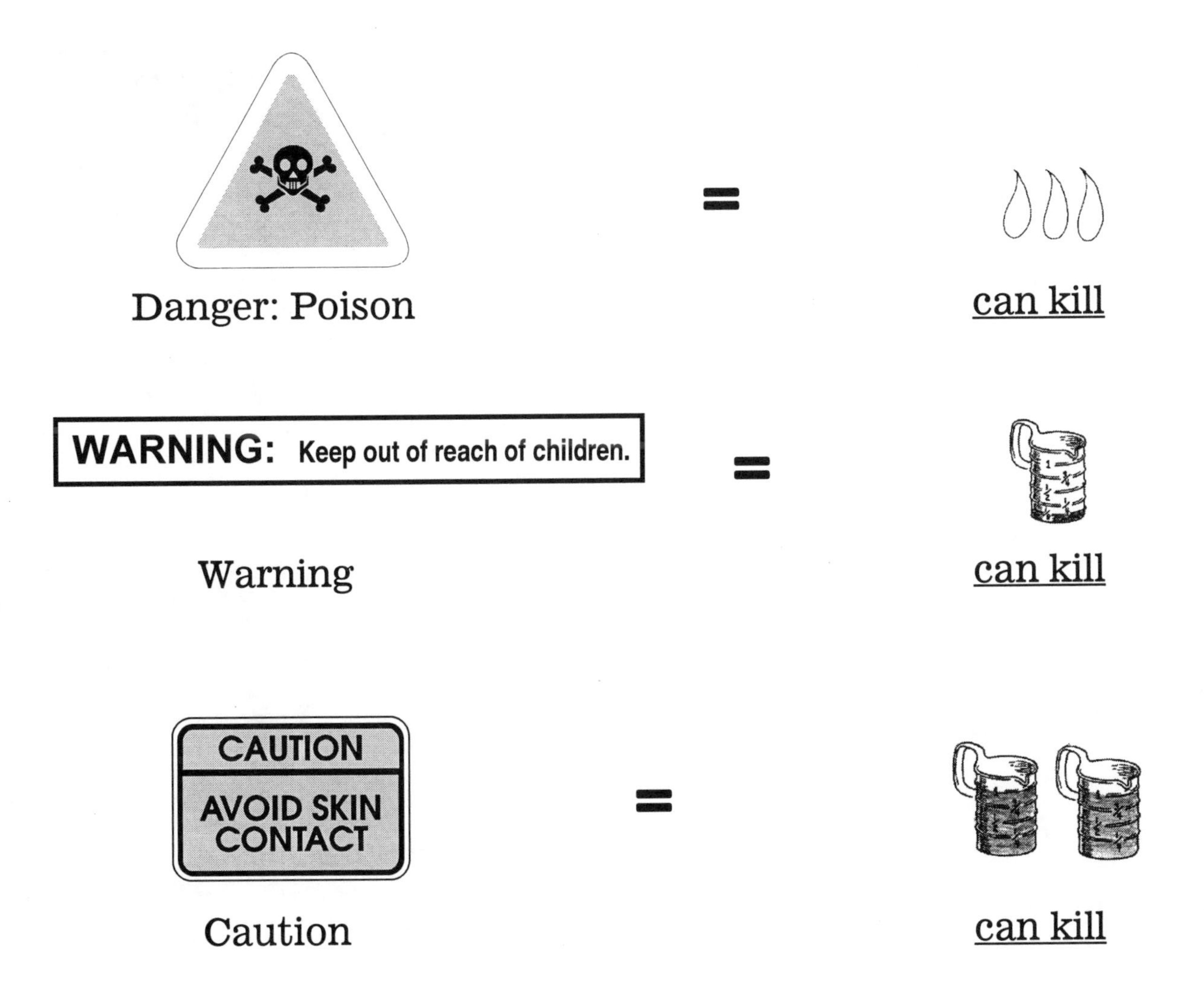

Danger: Poison = can kill

Warning = can kill

Caution = can kill

Circle the words you know.

Where do you see this label?

This pesticide is an _____ cide. It kills ____________.

The name of this pesticide is ______________.

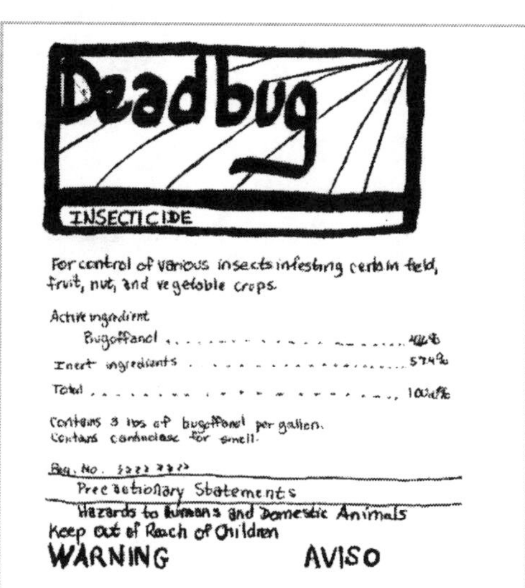

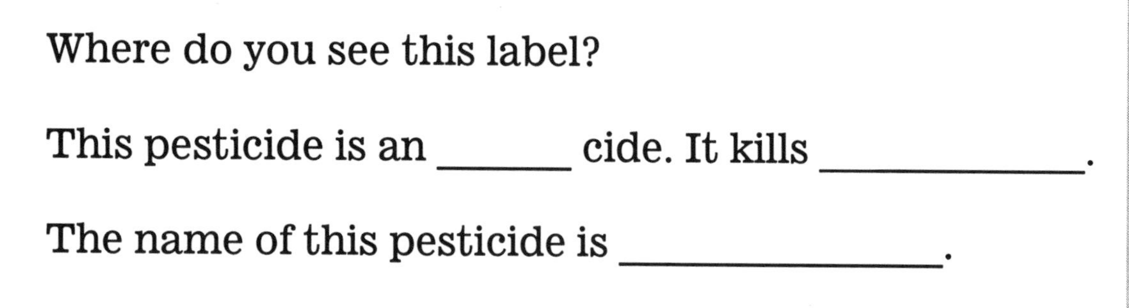

Where do you see this label?

This pesticide is an _____ cide. It kills ____________.

The name of this pesticide is ______________.

Moldaway

fungicide

Wettable Powder

Active Ingredient	By Weight
Moldnil	50%
Inert Ingredients	50%
TOTAL	100%

KEEP OUT OF REACH OF CHILDREN

CAUTION – PRECAUCIÓN

HAZARDS TO HUMANS AND DOMESTIC ANIMALS

Where do you see this label?

This pesticide is a ________ cide. It kills ____________.

The name of this pesticide is _______________.

Which pesticide is the most dangerous?_____________________________.

Keep pesticides from entering your body.

Do not get pesticides in your eyes.

Do not breathe in pesticides.

Do not swallow pesticides.

Do not get pesticides on your skin.

Pesticides can be...

liquids or sprays

powders or granules

gases

Which pesticides do you see at work?

Make an X under the word or words below.

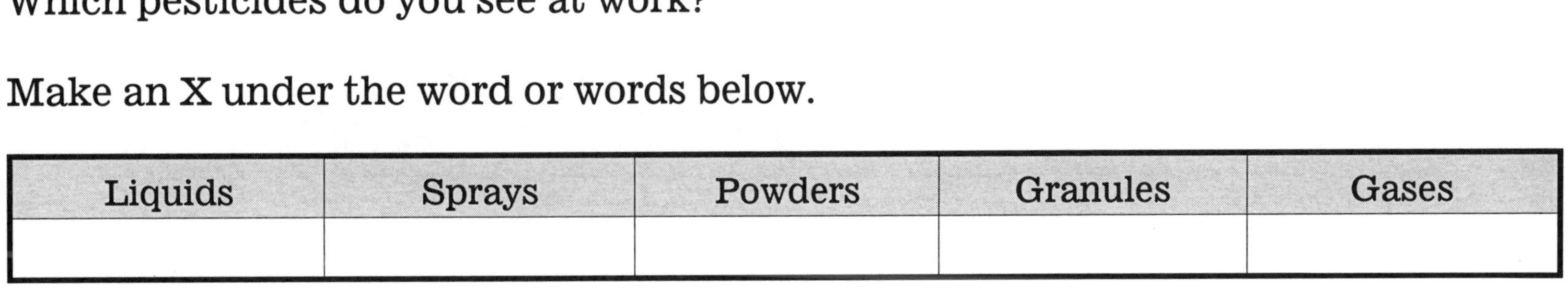

Liquids	Sprays	Powders	Granules	Gases

You can find pesticides...

on plants

in the soil

You can find pesticides...

in irrigation water

on irrigation machinery

You can find pesticides...

in storage places

on application equipment

Pesticides can drift.

Which pesticides do you see at work? Ask your boss.

Pesticide Name	Caution	Warning	Danger
WeedKill			X

Do you have any pesticides or anything else toxic in your house?

What is it?

What word of warning is on the label?

Lesson 2
Review

Check the correct answers.

1 How can pesticides get into your body?

_____ skin _____ eyes _____ smoking

_____ eating _____ drinking

_____ breathing _____ swallowing

2. Where can you find pesticides at work?

_____ on plants _____ on machinery _____ in irrigation water

3 Lesson Three

How can pesticides hurt me?

Pesticides are dangerous. They can make you very sick.

Sometimes you can get sick immediately.

Sometimes you may get sick later.

Pesticides can give you cancer or hurt your body in other ways.

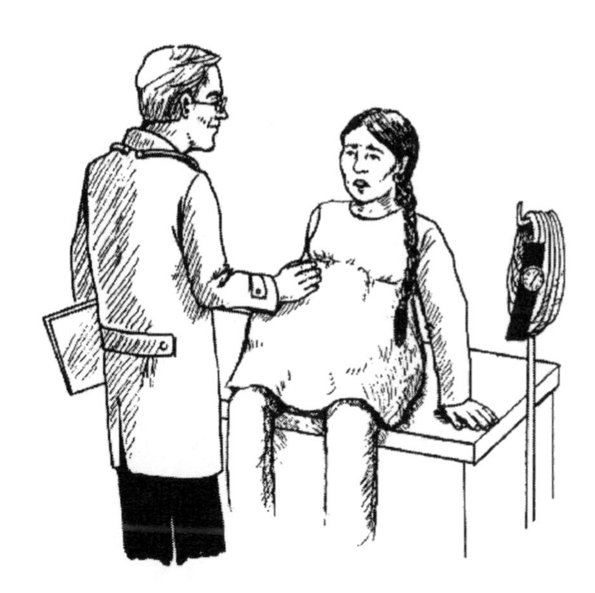

Pesticides can hurt your unborn baby.

Symptoms of Pesticide Poisoning

Doctor:	Do you have a rash?
Juan:	Yes, I do.
Doctor:	Do you have a headache?
Juan:	No, I don't.

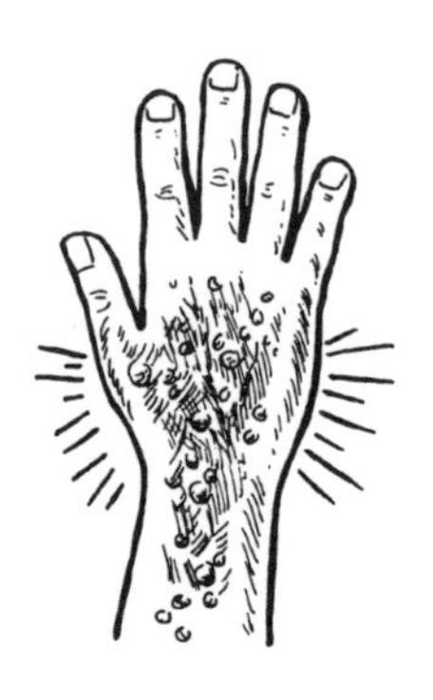

a rash

a headache

muscle pain

cramps

Fill in the blanks.

1. Does he have cramps?

   No, ______________ ______________.

   He has ______________ ______________.

2. Does she have a rash?

   Yes, ______________ ______________.

3. Do they have headaches?

   No,______________ ______________.

   ______________ ______________ ______________.

4. Do you have muscle pain?

   ______________, I ______________.

More Symptoms

Doctor:	Are you sweaty?
Carlos:	Yes, I am.
Doctor:	Are you dizzy?
Carlos:	No, I'm not.

sweaty

dizzy

weak

Fill in the blanks.

1. Is he sweaty?

   No, _______________ _______________.

   _______________ _______________.

2. Is she dizzy?

   Yes, _______________ _______________.

3. Are they weak?

   Yes, _______________ _______________.

4. Are you weak?

   _______________, _______________ _______________.

He is...

drooling

vomiting

He has...

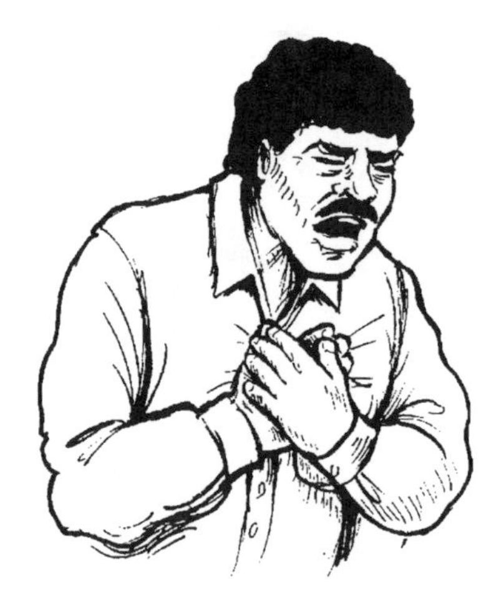

trouble breathing

small pupils

Fill in the blanks using am, is, are, has, or have.

1. She _______________ weak.

2. You _______________ a rash.

3. They _______________ trouble breathing.

4. He _______________ dizzy.

5. I _______________ vomiting.

6. She _______________ small pupils.

7. You _______________ drooling.

8. He _______________ a headache.

What's the matter?

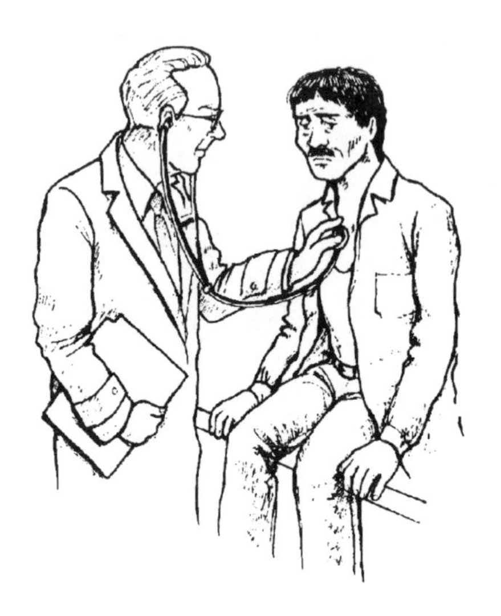

Doctor:	What's the matter?
Manuel:	I'm weak and dizzy and I have a bad headache.
Doctor:	Do you have muscle pain and cramps?
Manuel:	Yes, I do.

1. What's the matter with Manuel?
2. Does Manuel have a rash?
3. Does Manuel have muscle pain?
4. Does Manuel have cramps?

Match the word to the picture.

1. 
2. 
3. 
4. 
5. 
6. 
7. 
8. 
9. 
10. 
11. 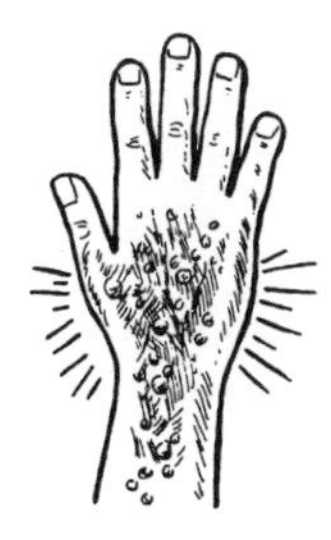

sweaty _____ headache _____ rash _____ trouble breathing _____

weak _____ cramps _____ muscle pain _____ vomiting _____

drooling _____ small pupils _____ dizzy _____

Symptoms Bingo

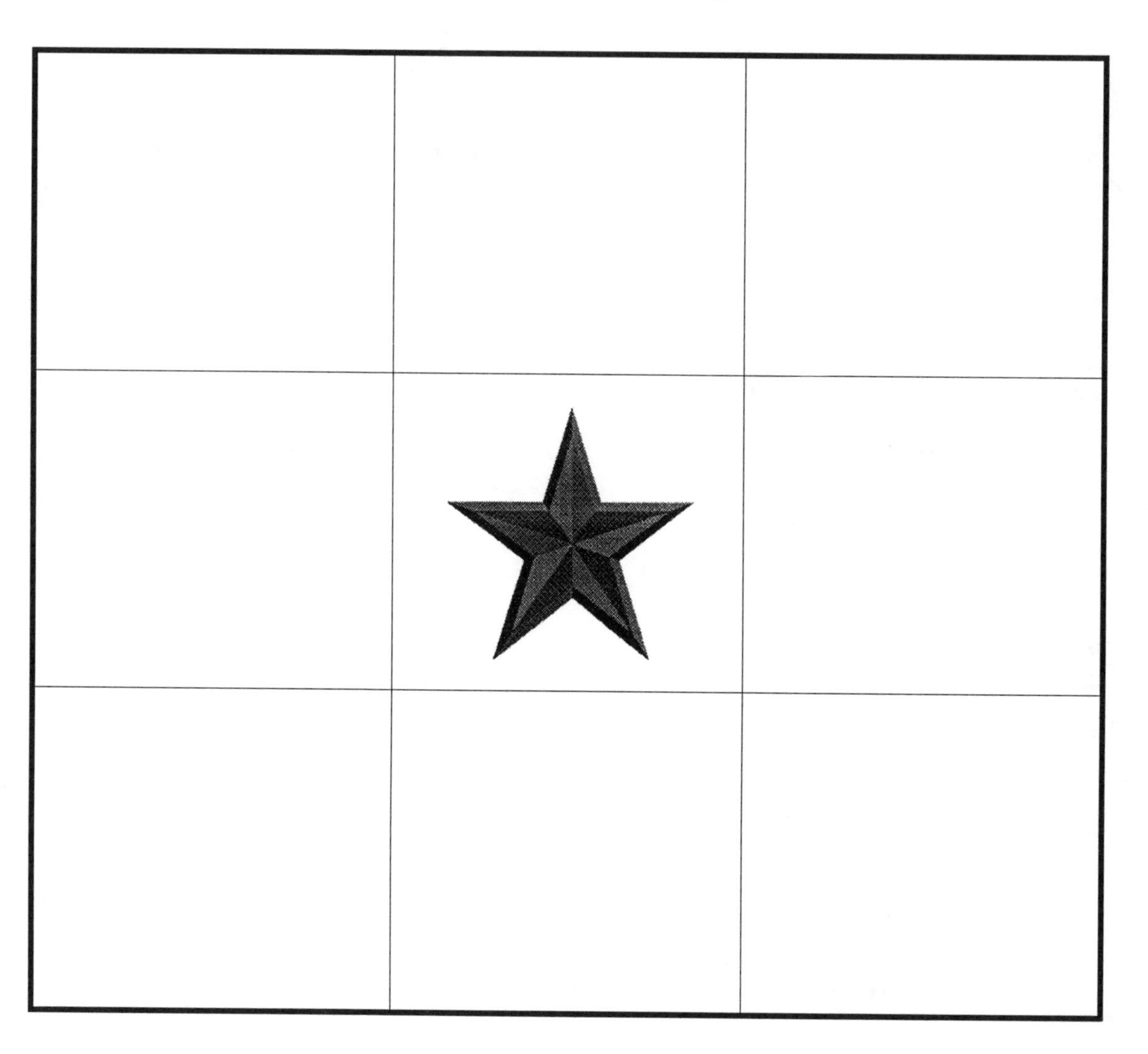

Lesson 3
Review

Check the correct answers.

1. Pesticides can make you sick...

immediately

later

2 What are the symptoms of pesticide poisoning?

vomiting

muscle pain

sweating

headache

4

Lesson Four

If I'm sick at work, what should I do?

Where can you get medical help?

Nearest medical center

Name ______________________________

Street ______________________________

City ______________________________

☎ Telephone ____________________

Where is the nearest telephone? ____________________

Emergency Plan for Medical Help

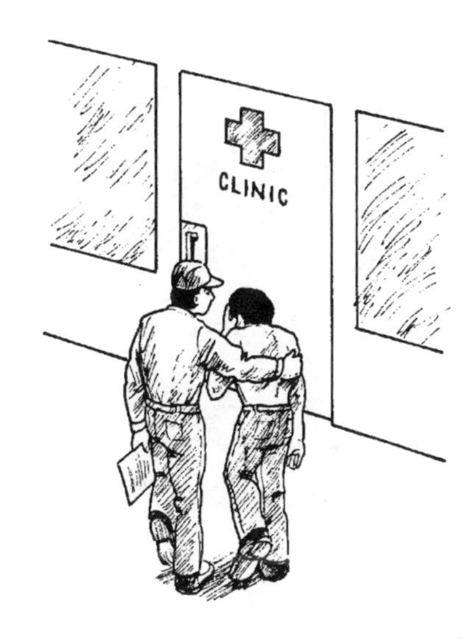

Answer these questions.

Where is the nearest doctor? ______________________________

When is the clinic open? ______________________________

How will you get there? ______________________________

What are the directions? ______________________________

What will you tell the doctor? ______________________________

If you get sick at work, tell your boss immediately.

Then, go to the doctor immediately.

Your boss must tell the doctor the name of the pesticide that made you sick.

You have pesticide in your eye. What should you do?

1. Hold your eye open. Rinse your eye with cool water. Rinse for 15 minutes.

2. Go to the doctor.

You have pesticide on your skin. What should you do?

1. Take off your clothes.

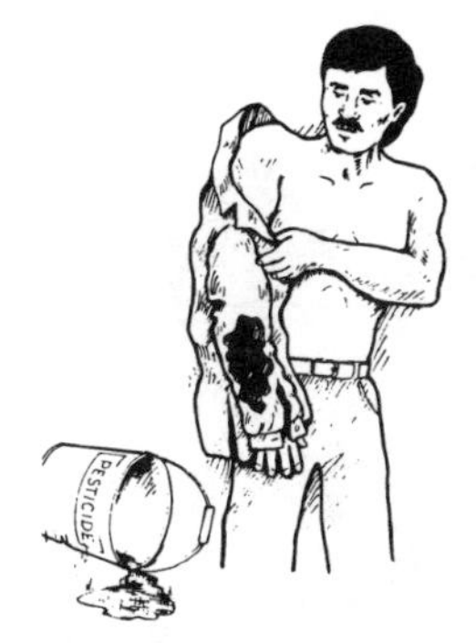

2. Rinse your skin with water immediately.

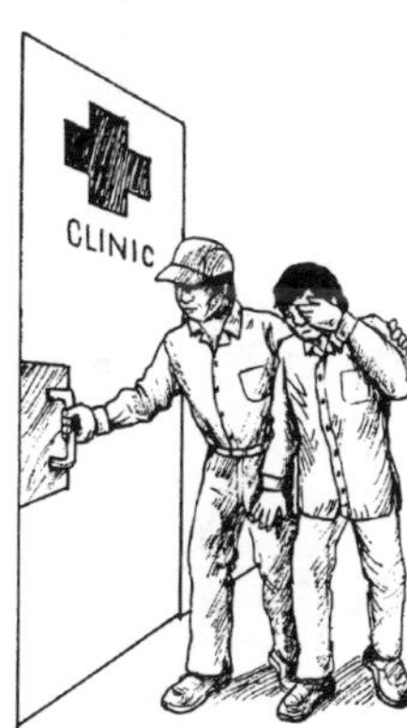

3. Wash with soap and water as soon as possible.

He swallowed a pesticide. What should she do?

1. Call 911, the poison control center, or go to the doctor.
   Give the name of the pesticide.
2. Read the first aid directions on the label. Follow the directions.
3. Go to the doctor as soon as possible.

Poison Control Center

☎ Telephone ______________________________

Street ______________________________

City ______________________________

You feel sick and dizzy in the greenhouse. What should you do?

1. Leave the greenhouse immediately.

2. Breathe fresh air.

She got sick from breathing pesticides. What should he do?

1. Get her to fresh air immediately.

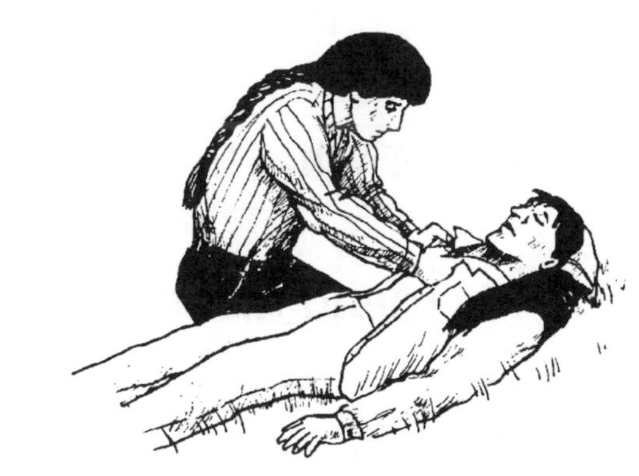

2. Loosen her clothes.

3. The woman is not breathing.

   Give mouth-to-mouth.

4 Call 911.

Calling 911

911 Operator: Hello. What is the nature of your emergency?

Florinda: My friend just swallowed a pesticide.

911 Operator: What is your address?

Florinda: The camp on Route 234. It's two miles from Greensboro Road.

911 Operator: Do you know the pesticide name?

Florinda: Yes. It's WeedKill.

He passed out in the greenhouse. What should you do?

1. Get help.
2. Do not go in!

You must have special breathing equipment and training to go into the greenhouse.

Lesson 4
Review

Number the pictures.
You have pesticide on your skin. What should you do?

______

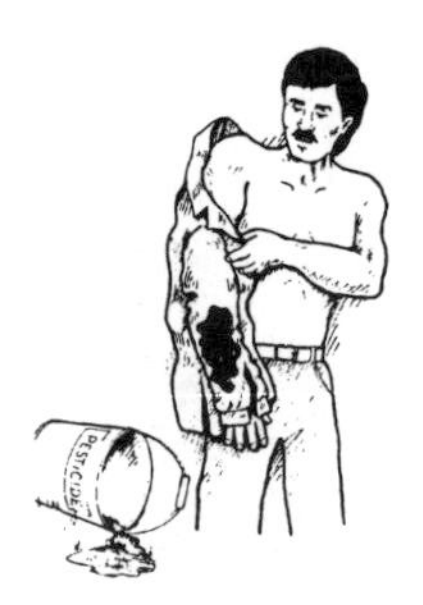
______

______

Check the correct answers.
You get sick at work. Whom do you tell?

______

______

______

5 Lesson Five

How can I protect myself from pesticides?

1. Wear clothes that cover your skin.

2. Look for soap, water, and towels.

3. Wash your hands before you eat, drink, smoke, chew gum or tobacco.

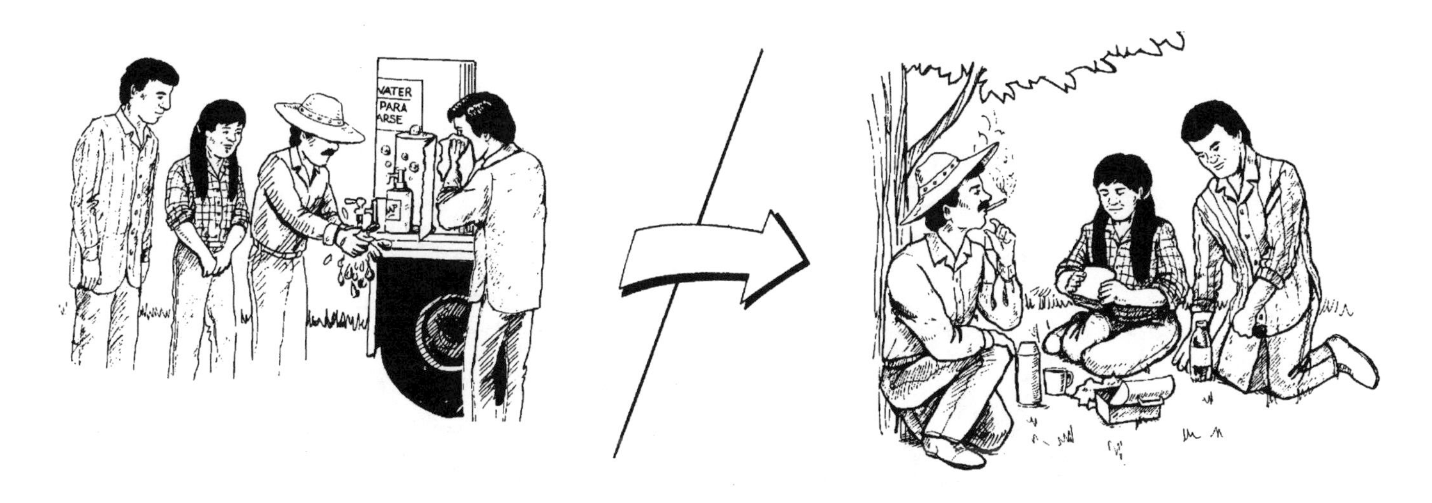

4. Wash your hands before you use the toilet.

5. When people are spraying pesticides, stay away!

Pesticides can drift.

6. A danger sign means keep out.

7. Do not enter areas when the boss says, "Stay out!"

8. Don't take pesticide containers home.

Don't use empty pesticide containers.

9. Keep children away from pesticides.

10. Wash your body and hair after work each day.

Use soap. Put on clean clothes.

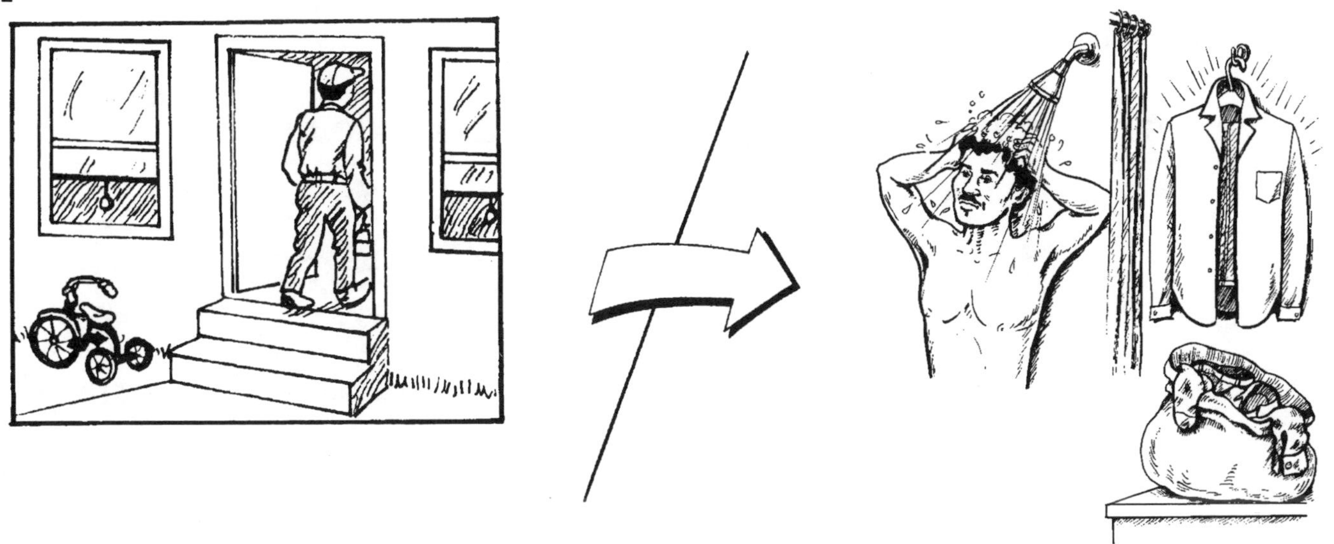

11. Keep dirty work clothes away from non-work clothes.

Change your work clothes every day.

Lesson 5
Review

Check the correct answers.

1. When you arrive home from work, what should you do first?

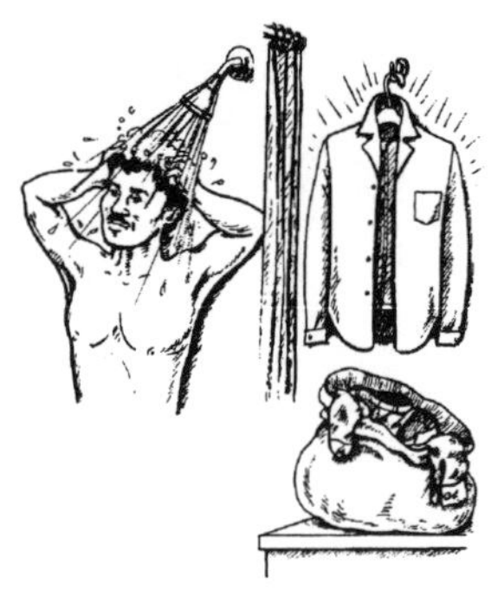

______ ______

2. If pesticides drift into your work area, what should you do?

______ ______ ______

3. Where should you put empty pesticide containers?

______

______

______

Pesticide Safety Game

Player A:
Toss a coin.

Look at the words in Box A. Is it there?

Yes: Put an x next to the rule.

No: Player B tosses the coin.

Box A:
Look for soap, water, and towels.
Wash your hands before you use the toilet.
When people are spraying pesticides, stay away!
A sign means keep out.
Keep children away from pesticides.
Keep dirty clothes away from clean clothes.

Player B:
Toss the coin.

Look at the words in Box B. Is it there?

Yes: Put an x next to the rule.

No: Player A tosses the coin.

Box B:
Wear clothes that cover your skin.
Wash your hands before you eat, drink, smoke, chew gum or tobacco.
Don't enter areas when the boss says, "Stay out."
Don't take pesticide containers home.
Wash your body and hair after work.

The winner has the most x's.

Pesticide Safety Game

Start

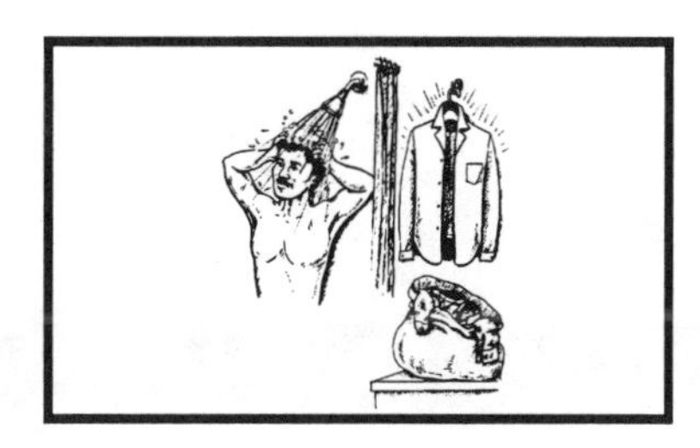

Finish

Make a warning sign.

6

Lesson Six

What are my rights?

Your boss must tell you about pesticide use at work.

Fill in the pesticide application information.

Pesticide Application

Where: ______________________________

What: ______________________________

When: ______________________________

Do not enter until: ____________________

Certron 5 7/29/97 7/27/97 apple orchard

Your boss must have soap, water, and towels near your workplace.

Your boss cannot make you work in dangerous areas.

Dangerous areas are where

workers are putting pesticides on the crops

and pesticides can drift onto you

or under a restricted entry interval.

Find the Restricted Entry Interval (REI).

Pesticide	REI
Certron 5	24 hours
Mixchlar	12 hours
Rothione	48 hours

Which pesticide is the most dangerous?

Which pesticide is the least dangerous?

For some jobs, the boss must give you special training and protection.

A pesticide handler, flagger, or early entry worker must have special training.

The boss must not punish you for following rules.

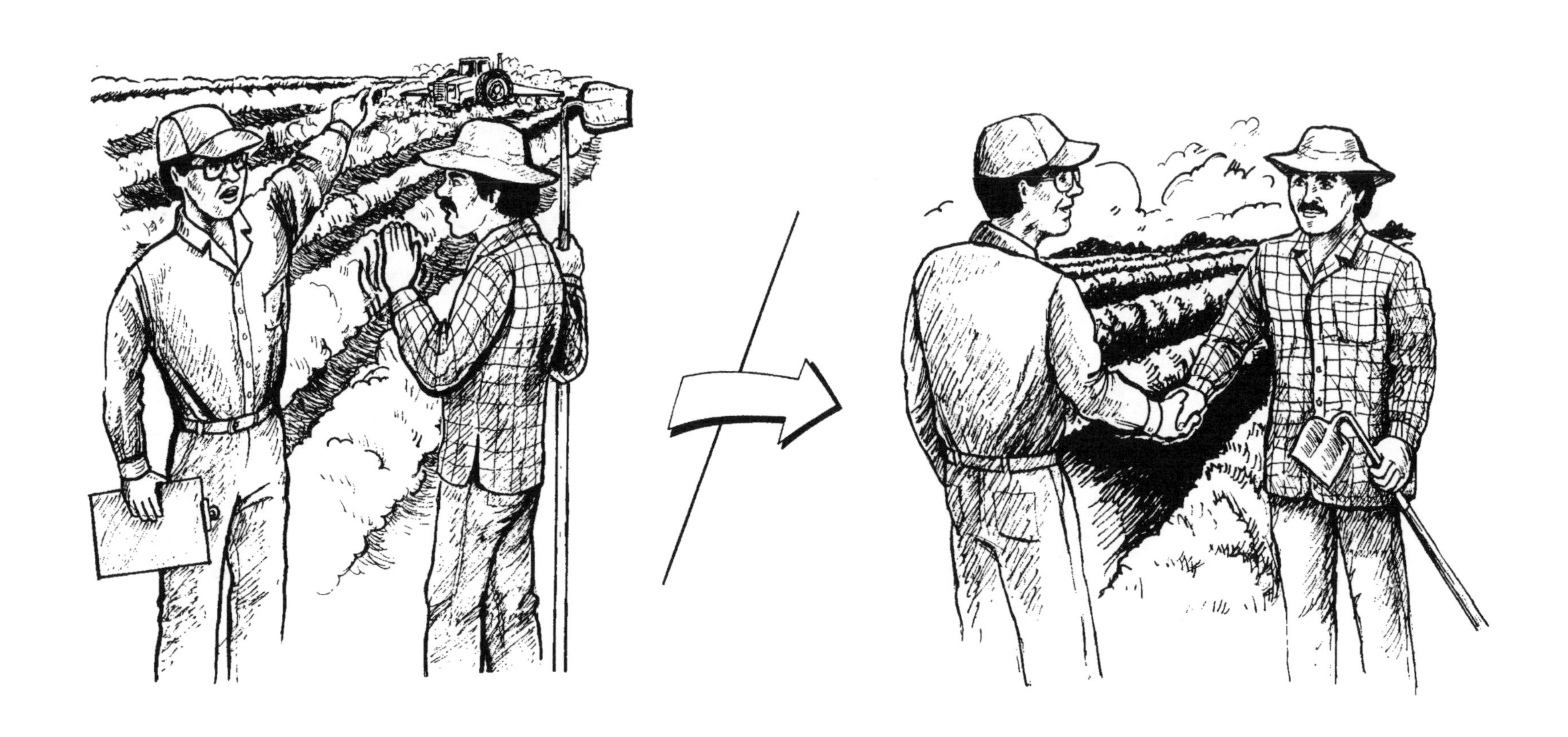

Get to work!

"Get to work!"

"I can't now, it's dangerous."

Answer the questions.

1. Who are the men? ____________________________

2. What does the boss want? ____________________________

3. What does the worker want? ____________________________

4. What happens next? ____________________________

Lesson 6
Review

Check one.

1. Can your boss fire you for following pesticide safety rules?

yes ______

no ______

2. Must your boss tell you about pesticide use at work?

yes ______

no ______

3. Must your boss give you special training and protection for these jobs?

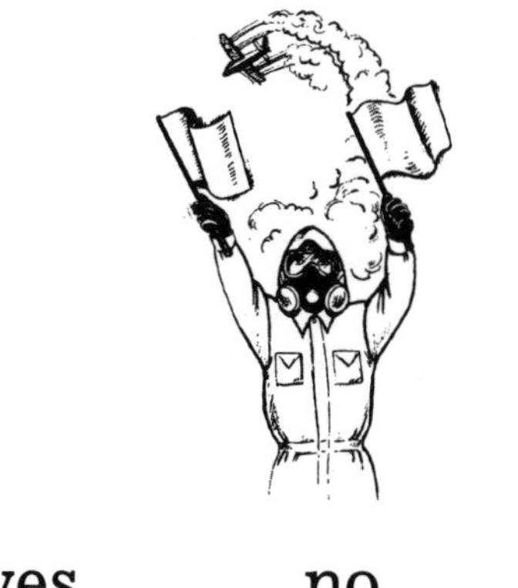

yes _____ no_____

yes_____ no _____

Lesson Seven

Be safe on the farm.

Be safe in the sun.
Don't get a sunburn.

Wear a hat. Cover your arms and legs.

Be safe in the heat.
Don't get heat stress.

Drink many glasses of cool water.

Take breaks often.

Don't hurt your back. Do your work safely.

Stretch before you lift.

Lift with your legs.

Do not lift with your back.

Get help to lift heavy things.

Make your stomach muscles strong.

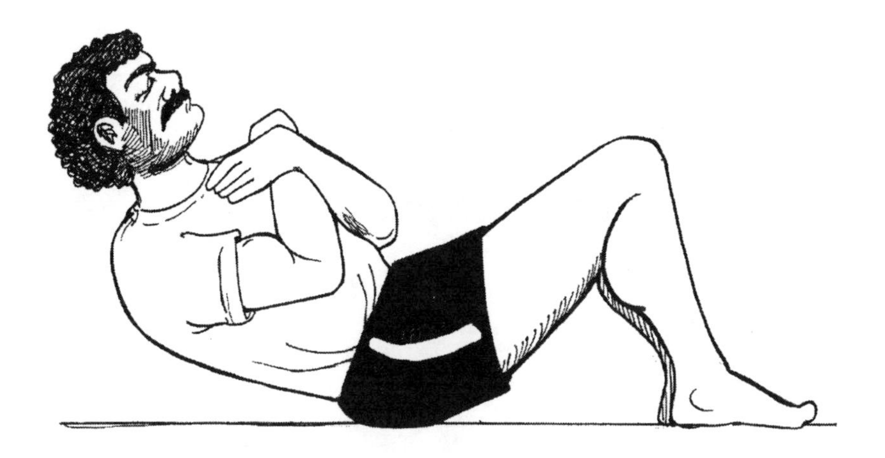

Do you work close to the ground?

Get down on one or two knees to work.

Remember to stand and stretch.

Do you have to stand for a long time?

Stand with one foot on a box.
After a while, put your other foot on the box.

Wear comfortable shoes.

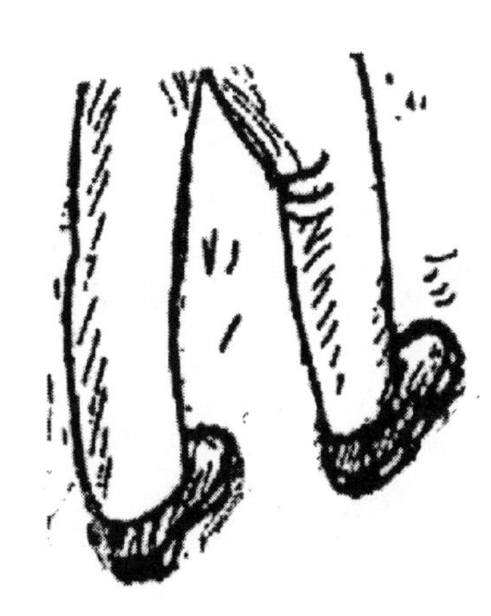

Do you sit or drive for a long time?

Sit up straight.
Put a small cushion behind your lower back.

Take a break to stand and stretch.

Do your work safely.

Don't have an accident.

Look out for dangerous jobs.

Tell other workers about possible dangers.

Be careful with machines.

Turn off equipment before fixing problems.

Do you have children?

Do not let a child ride on a tractor with you.

Do not let children go near dangerous equipment.

Look out for children playing near you.

Talk to your children about farm safety.

Draw an accident.

Repeating the same motion can hurt you.

Take a break. Stretch, rotate, and shake your hands.

Use safe equipment.

Don't drink or wash in irrigation water.

Make sure well water is safe to drink.

Lesson 7
Review

Check the correct answer.

1. Which is the correct way to lift things?

______ ______

2. Which are the correct clothes to wear in the sun?

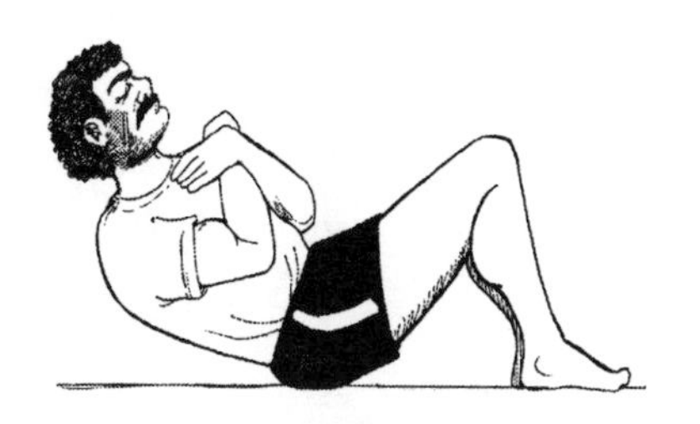

______ ______